Merrow Technical Library
Practical Science
STRAIN GAUGES

Merrow Technical Library

General Editor: J. Gordon Cook, B.SC., PH.D., F.R.I.C.

Practical Science

STRAIN GAUGES

E. J. Hearn, B.Sc.(Eng.) Hons., C.Eng., M.I.Mech.E.
*Senior Lecturer, City of Birmingham Polytechnic,
Birmingham, England.*

MERROW

Merrow Publishing Co. Ltd.
276, Hempstead Road,
Watford   Herts   England

© 1971 *Merrow Publishing Co. Ltd.*

All rights reserved. No part of this publication may be reproduced, stored in a retrieval system, or transmitted, in any form or by any means, electronic, mechanical, photocopying, recording or otherwise, without the prior permission of the publishers

ISBN 0 900 54124 5 Clothbound
ISBN 0 900 54125 3 Paperback

Printed in Great Britain at the Pitman Press, Bath

CONTENTS

1	Introduction	1
2	Basic Principles	3
3	Gauge Selection	10
4	Factors Influencing Gauge Selection and Performance	18
5	Strain Gauge Installation	26
6	Gauge Protection	34
7	Effect of Strain Gauge Excitation Current	37
8	High-temperature Measurements	40
9	Long-term Testing	43
10	Dynamic Measurements from Rotating Components	46
11	Basic Measurement Systems	49
12	D.C. and A.C. Systems	54
13	Recording Systems	58
14	Other Types of Strain Gauge	59
15	Advantages of Electrical-resistance Strain Gauge Technique in Experimental Stress Analysis	65
	Bibliography	67
	Appendix	
	Firms Manufacturing Strain Gauges, Strain Gauge Instrumentation, Ancillary Equipment and Associated Recording Equipment	69
	Centres from which Advice on Strain Gauges may be Obtained	72
	Index	73

1

INTRODUCTION

The accurate assessment of stresses and strains in components under working conditions is an essential requirement of engineering design. The location of principal stresses, peak stress values and stress concentrations, and their reduction or removal by suitable design, has applications in every field of engineering.

The electrical-resistance strain gauge was introduced in the U.S.A. in 1939 by Ruge and Simmons. Since that time, it has come into widespread use, particularly in the aircraft industry, and is now the basis of one of the most useful experimental stress-analysis techniques. The strain gauge has contributed in no small measure to the high level of safety and performance of modern-day aircraft.

Today, the use of strain gauges is by no means restricted to the aircraft industry; strain-gauge techniques have been applied to stress-analysis problems in almost every branch of engineering.

Strain gauges have been used for measurements on all manner of objects, ranging from nuclear boilers, turbine blades, vehicle engines and chassis, to human bones and skin. Strain gauges have measured the forces developed by a chick inside its egg; they have followed the forces involved in the act of human birth.

After 30 years of intensive research and development, electrical strain gauges have evolved into four main classes:

(a) Bonded wire gauges.
(b) Bonded foil gauges.
(c) Unbonded wire gauges.
(d) Piezo-resistive (semi-conductor) type gauges.

Bonded wire and foil gauges are now the most widely used, and the text of this monograph is concerned mainly with these types of gauge. They consist essentially of a length of wire or foil, suitably insulated, which is bonded to the surface of the component under test. When it is stretched or compressed under an applied strain, the gauge records a change in the electrical resistance of the wire or foil. With the help of suitable instrumentation, this resistance change can be calibrated to indicate strain; it thus provides a direct method of strain measurement.

The unbonded wire gauge, which was used mainly in transducer design, has been largely superseded by advanced foil gauges.

The semi-conductor gauge is used in special cases where the magnitude of the signal is such that considerable amplification is required; this type of gauge is described on page 60.

The relative simplicity of the strain-gauge technique, and the fact that it is applicable directly to components under service loading conditions, have made it the most popular and widely-used technique of experimental stress analysis.

2

BASIC PRINCIPLES

The fundamental equation for the electrical resistance R of a length of wire is

$$R = \frac{\rho L}{A}$$

where L = length, A = cross-sectional area, and ρ is a constant, the specific resistance or resistivity.

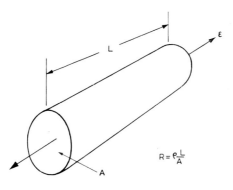

Fig. 1

If the wire is subjected to compression or tension, its length and hence its cross-sectional area will change; it follows, therefore, that its resistance will change too.

An electrical-resistance strain gauge is constructed from a grid of wire or foil, as shown in Fig. 2, bonded to a non-conductive backing; it is cemented firmly to the component

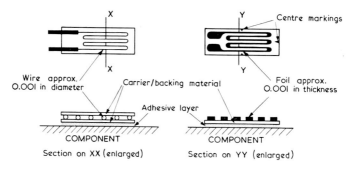

Fig. 2. Basic format of wire and foil gauges

whose strain is to be measured. Any slight strain on the component is transmitted to the gauge, which accordingly changes in length. This results in a small resistance change, and a sensitive and suitably-calibrated instrument records the change directly as strain.

Experiment shows that, for a number of alloys (notably those of copper and nickel) there is a direct relationship, over a considerable range of strain, between change of resistance and strain, as expressed by the following equation:

$$\frac{\Delta R}{R} = K \times \frac{\Delta L}{L} \qquad (1)$$

where ΔL and ΔR are the changes in length and resistance respectively. K is termed the gauge factor; it lies usually in the region of 2 to 2·2 for most conventional gauges.

Thus:

$$\text{gauge factor } K = \frac{\Delta R/R}{\Delta L/L} = \frac{\Delta R/R}{\varepsilon}$$

where ε is the strain.

The value of the gauge factor for a batch of strain gauges (obtained from calibration tests) is always supplied by a manufacturer. Most modern strain gauge instruments may be set according to this value, enabling strain to be recorded directly.

The resistance changes in strain gauges are of a very small order, and sensitive instrumentation is required. An appreciation of a typical value of ΔR can be obtained by considering the following example, in which it is required to record the stress in a steel bar to an accuracy of, say, 150 lbf/in^2.

Assuming a gauge factor of 2, a gauge resistance of 120 Ω and E for steel $= 30 \times 10^6$ lbf/in^2:

$$E = \frac{\sigma}{\varepsilon}$$

$\therefore$ Strain $\varepsilon = \dfrac{\sigma}{E} = \dfrac{150}{30 \times 10^6}$

$\qquad\qquad = 5 \times 10^{-6}$ i.e. 5 microstrain

From Eqn. (1) $\Delta R = K \times R \times \varepsilon$

$\qquad\qquad = 2 \times 120 \times 5 \times 10^{-6}$

$\qquad\qquad = \underline{0\cdot 0012 \text{ ohm.}}$

Strain-gauge instruments, or bridges as they are normally termed, are basically Wheatstone bridge networks as shown in Fig. 3. The condition of balance for this network is obtained when:

$$R_1 \times R_3 = R_2 \times R_4$$

and the galvanometer will read zero.

Unbalanced Bridge Circuit

In the simplest half-bridge wiring system, resistances R_3 and R_4 are matched within the instrument as precision resistors; R_1 is an active gauge, i.e. one mounted on the component to be strained, and R_2 is a second gauge (identical with R_1) mounted on a similar piece of material which remains unstrained throughout the test. R_2 is termed a "dummy" gauge,

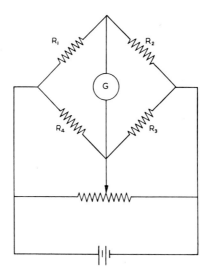

Fig. 3. Wheatstone bridge circuit

its function being to cancel out resistance changes due to temperature variations. R_1 and R_2, being similar gauges on "opposite sides" of the bridge, will change in resistance by the same amount as a result of temperature variation, and such changes will not alter the galvanometer reading.

With the bridge balanced initially on zero, any strain on gauge R_1 will cause the galvanometer to deflect; this deflection can be calibrated as strain by including in the circuit an arrangement whereby the gauge factor of the gauges used can be taken into account.

Null Balance or Balanced Bridge Circuit

In an alternative arrangement, a variable resistance can be included in one arm of the bridge, and any deflection of the galvanometer can be cancelled by suitable adjustment of the resistance. This adjustment can be calibrated directly to indicate strain. This is a null-balance or balanced-bridge circuit.

Grid Construction

The strain gauge may be constructed from wire or foil arranged in the form of a grid. This type of construction increases the sensitivity of the device.

If we consider, as an analogy, a length of hose pipe carrying a flow of water, the flow will change if the pipe is stretched; the frictional resistance of the walls of the pipe has been increased due to its increased length. Similarly, the flow of current in a wire will decrease when the wire is stretched, and a change in resistance is recorded. The sensitivity of both systems can be increased, however, if the hose and wire respectively are looped into a grid as shown in Fig. 4. Less

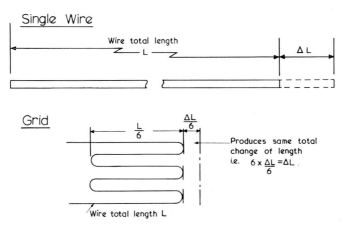

Fig. 4. Effect of looping wire into a grid

stretching is now necessary to bring about an equivalent change of overall length; a grid consisting of six sections, for example, will gain in overall length six times as much as a single hose or wire stretched by the same amount.

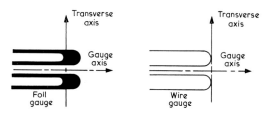

Fig. 5. Magnified views of the ends of foil and wire grids, showing significantly greater area for the foil gauge and therefore much lower resistance and transverse sensitivity

Transverse Sensitivity

At the end of each loop of wire in a grid there is a portion of wire aligned at right angles to the gauge axis. This will be sensitive to strains perpendicular to the gauge axis, and any such strains will produce a change in resistance of the gauge in addition to the required signal. This effect is termed the *transverse sensitivity* or *cross sensitivity*[26] of the gauge; except in a particular case,* it has a detrimental effect on gauge performance.

Transverse sensitivities are normally much higher in wire than in foil gauges; in the latter, the area is deliberately increased at the end, as shown in Fig. 5, to reduce the cross-resistance and hence the cross-sensitivity. Cross-sensitivities of single-element foil gauges are commonly less than 2 per cent, and they have negligible effect on the gauge output.

Wire or Foil?

Selection of the type of gauge to be used is influenced largely by the factors of cost and cross-sensitivity. Wire gauges are, in general, cheaper than foil gauges, but the latter have lower cross-sensitivity.

Foil gauges, however, offer other advantages over wire gauges, as follows:

* See "Stress Gauge", page 16.

(a) Very thin foil gauges are available, with a high degree of flexibility; they require no correction in bending applications, i.e. they are effectively in the surface itself.
(b) Grid elements in foil gauges are significantly wider than their thickness, producing relatively large bonding areas. Creep and hysteresis losses are subsequently much lower.
(c) Foil gauges display from 5 to 20 per cent greater sensitivity than equivalent bonded wire gauges.
(d) The range of types of foil gauge available is significantly greater than the range of wire gauges, particularly in respect of self-temperature-compensated gauges.

Much of the development work in strain gauges is concerned with foil gauges, and very high precision gauges of this type are available. In most general applications, however, the author has found either type of gauge to give acceptable results, provided adequate care is taken in the installation process.

3

GAUGE SELECTION

There are many different types, forms and sizes of strain gauge, each one being designed for particular applications. Examples of the wide range available are described in this section.

Linear Gauges

Linear gauges are the most widely-used form of gauge; they are designed to record the strain along one particular axis. Figure 6 shows a linear gauge of foil construction. Markings are included in the format of the gauge to indicate precisely the centre of the active part of the gauge, i.e. the gauge length. This may vary from a few thousandths of an inch to several inches[27], the length selected depending on the application and the rate of change of strain at the point under consideration. In a fairly uniform strain field, a suitable gauge length would probably be $\frac{1}{4}$ or $\frac{1}{2}$ inch.

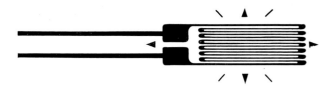

Fig. 6. Typical linear foil gauge showing centre position markings and increased element area at the end of each loop

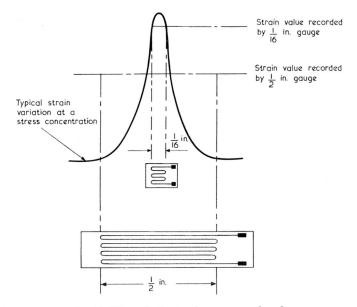

Fig. 7. Effect of using too large a gauge length

Great care must be taken in choosing the gauge length for a particular application. It must be remembered that the strain gauge will record the average value of strain across its length. At sharp stress concentrations, therefore, the usual ½ inch gauge might well be too large; Fig. 7, for example, shows a typical strain variation at a stress concentration. Here, the strain value recorded by the ½ inch gauge is considerably lower than the peak value which would be shown by, say, a $\frac{1}{16}$ inch or smaller gauge.

On the other hand, it is necessary to ensure that the gauge selected is not too small. Where overall strain values are required in a concrete structure, for example, small gauges might record only the value of local strains at inclusions, etc. In this case, gauges of at least four times the aggregate size should be used. This is a general rule that should be followed in using strain gauges with all non-homogeneous materials.

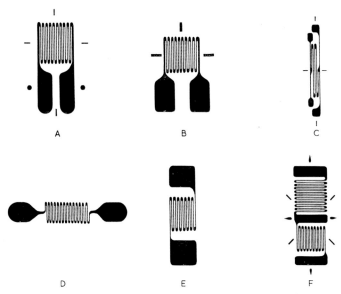

Fig. 8. Typical foil gauges. A: Single-element strain gauge. B: Single element gauge with large tabs for easy soldering. C: Double-ended gauge (tabs on both ends) for minimum width applications. D: Strain gauge for minimum length applications when width is available. E: Strain gauge for minimum width applications. F: Two-gauge rosette for simple tension or compression type load cells
(Brush Instruments.)

In general, it is advisable to use as large a gauge as possible, with high resistance values giving strong ouput signals less likely to be influenced by spurious noise signals, etc.

Rosette Gauges

In certain cases, e.g. in complicated components or loading conditions, the direction of the greatest (or principal) strain is not always known, and a single linear gauge cannot be used. In such cases, three or more gauges at known orientation with respect to each other, together forming a rosette gauge, are bonded at the point. The readings of the gauges

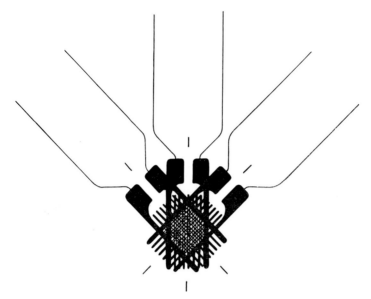

Fig. 9. 'Stacked' rectangular rosette. Gauges stacked on top of each other to reduce overall size and locate more precisely the centre of the gauge.

can be converted to yield the magnitude and direction of the maximum strain.

Common types of rosette gauge are, (a) the rectangular rosette with gauges at 0°, 45° and 90° to the gauge axis, and (b) the delta rosette with gauges at 0°, 60° and 120° (see Figs. 9 and 10).

Torque Gauges

The principal stresses in a shaft subjected to torsion occur at 45° to the axis of the shaft. Figure 11 shows an arrangement of gauges used for recording torque or shear-stress values.

The grid elements are aligned at 45° to each other on each side of a central axis, on a common backing. The central axis is then arranged in such a way as to lie circumferentially around the shaft when the gauge is cemented in position. This is simpler and more accurate than attempting to bond single linear gauges at a 45° angle on a curved surface.

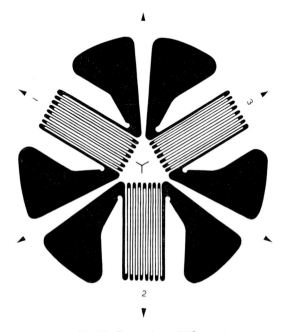

Fig. 10. Conventional 120° rosette

"Postage Stamp" Gauges

Gauges are available in which the backing (e.g. epoxy/cellulose) is impregnated with a water-sensitive adhesive. They may be moistened and pressed into place like postage stamps. Leads can be attached immediately. Drying times vary from 6 to 24 hours depending on temperature and humidity conditions; a heat lamp may be used, reducing drying time to as little as one hour.

Friction Gauges

The friction gauge is not bonded to the surface of the component under test; it relies upon friction to transfer strain to the gauge element.

The gauge is held on the surface with a magnetic pod or similar device. This type of gauge is particularly useful in surveying large areas to locate highly-stressed regions.

Self-adhesive Gauges[16]

The Hickson self-adhesive gauge may be used as an alternative to the friction gauge in many applications. Hickson gauges make use of the adhesion developed when a plate-polished sheet of elastomer is placed in contact with any other polished surface.

The metal filament of the gauge is vacuum-deposited on a thin polyester sheet and embedded in an elastomeric matrix. The inherent low stiffness of the assembly makes it particularly useful for application to thin sheet structures or low-modulus materials. The bond has a high shear strength and low peel strength, and the gauge can be applied and removed many times without deterioration.

The high temperature coefficient of the element precludes its use for static work, except under temperature-controlled conditions.

Fig. 11. Torque gauge

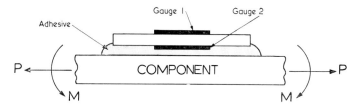

Fig. 12. 'Flexagauge' installation

Flexagauge

In certain applications, plane stresses must be distinguished from bending stresses. This is achieved by using a specially-prepared gauge which incorporates two gauges placed back-to-back on a piece of material of known thickness, as shown in Fig. 12.

Stress Gauge[21]

Transverse sensitivity is put to good use in the production of a strain gauge which has an output proportional to the *stress* in the direction of the gauge axis, rather than the *strain*. The grid of this so-called "stress gauge" is designed especially to produce the necessary transverse sensitivity.

Fatigue-Life Gauge

A gauge is available in which the electrical resistance changes in accordance with its fatigue experience. Changes of the order of 10 per cent are often obtained; this is several times larger than the resistance-change produced in equivalent strain gauges under normal levels of loading. Fatigue-life gauges are not strain gauges in the strict sense of the term.

Mounted on a structure or component at a point where fatigue cracks are expected, the fatigue-life gauge monitors resistance change which has been previously calibrated against fatigue damage for the material under test.

Measurement of fatigue damage can be made at intervals throughout the service life of components, using simple instruments which are connected to the gauge only for a period long enough to measure its resistance. Warning of potential fatigue failure, previously unobtainable by any other means, is thus obtained.

4

FACTORS INFLUENCING GAUGE SELECTION AND PERFORMANCE

The following factors play a significant part in the selection of gauges for particular applications, and have an important bearing on the performance of the installation.
   (a) Duration and frequency of strain.
   (b) Physical size and form.
   (c) Transverse sensitivity.
   (d) Operating temperature.
   (e) Strain limits and hysteresis.
   (f) Cost.
   (g) Fatigue life.
   (h) Gauge resistance and sensitivity.
   (i) Temperature compensation.
   (j) Method of installation and protection.

(a) Duration and Frequency of Strain

There are three main categories into which strain-gauge measurements can be classified. First, there is the measurement of static or very-low-frequency strains. For these purposes, conventional, relatively inexpensive gauges are adequate if sufficient care is taken with temperature compensation and protection of the gauge against hostile environments, e.g. water, oil, brine, etc. Self-temperature-compensating gauges are often ideally suited for this application (see page 22).

The second category includes the measurement of medium- and high-frequency dynamic strains, which may be of considerable magnitude. Under these circumstances, particular consideration should be given to the fatigue characteristics of the gauge element.

The third category includes measurement of combinations of static and dynamic strains, commonly under changing temperature conditions. The temperature-compensated gauge is often the preferred choice.

(b) **Physical Size and Form**

There is practically no limit to the shape or size of foil gauge which can be produced. The standard range available is so extensive that it is difficult to visualize any application which is not already catered for.

There is occasionally a need for very thin flexible gauges which can be moulded easily to difficult contours. This is achieved in practice by using, for example, foil gauges with very thin fibreglass laminates and epoxy-base carrier material.

(c) **Transverse Sensitivity**

This has been considered on page 8.

(d) **Operating Temperature**

Gauges are available which can operate from $-150°C$ to $+850°C$ for static measurements, and at even higher temperatures for dynamic applications. (No single gauge covers the whole range.) At extremes of temperature, however, the accuracy of results is debatable; a realistic upper limit for results which can be accepted with confidence might well be set in many cases as low as $300°C$. The development of reliable high-temperature strain measurements is a branch of strain-gauge operation which continues to present serious problems.

Basically, the operating temperature of a gauge is determined by the thermal characteristics of the gauge element material and the carrier material. Upper temperature limits for typical carrier materials are:

Paper	60°C
Epoxy Resin	100°C
Polyester	100–200°C
Glass Fibre	250°C
Phenolic Resin	250°C
Ceramic	400°C and above.

Special gauges and installation methods for use at higher temperatures are described on page 40 under *High Temperature Measurements*.

(e) **Strain Limits and Hysteresis**

The majority of conventional strain gauges will operate successfully up to 1 per cent strain (10,000 $\mu\varepsilon$). This is well beyond the elastic limit of most engineering materials. Accurate measurement of up to 5 per cent strain is possible by using appropriate standard self-temperature-compensated gauges, and special high-elongation gauges are available for strains up to 15 per cent elongation. Strain limits, therefore, seldom present a problem.

Combined hysteresis and zero shift tests with foil gauges during typical strain reversals of $\pm 1500\ \mu\varepsilon$ show errors induced to be of the order of less than 0·1 per cent; they may, in general, be ignored.

(f) **Cost**

The cost of strain gauges varies greatly depending on type, size, form, manufacturer and application. In the following list, types of strain gauge are arranged in order of increasing cost. The arrangement serves only as a general guide, however, as the costs of individual installations depend upon specific requirements.

1. Conventional non-temperature-compensated wire gauges.
2. Conventional non-temperature-compensated foil gauges.

3. Temperature-compensated foil gauges.
4. Rosette gauges.
5. High-temperature gauges.
6. Self-adhesive gauges.
7. Semi-conductor gauges.

(g) Fatigue Life

All metals are subject to fatigue damage under conditions of reversed stress at sufficiently high levels, and those used as elements for strain gauges are no exception.

Fatigue damage in strain gauges produces a permanent change in the unstrained resistance of the gauge, i.e. a "zero shift". In some cases, e.g. in dynamic strain measurement, zero shift does not present a problem; the gauge can be considered satisfactory up to the point where cracking of the element has occurred.

In static or combined static/dynamic measurements, zero shift may create a considerable problem, and the gauge must be regarded as inadequate if the shift is greater than 5–7 per cent of the total strain range.

Fatigue information available from manufacturers has been obtained usually from fatigue tests carried out under a $\pm 1500\,\mu\varepsilon$ reversal; in such circumstances, a fatigue life in excess of 10^6 cycles should be expected from good foil gauges. In many cases, 10^7 is achieved.

Increase of the strain level to, say, $\pm 2{,}000\,\mu\varepsilon$ can reduce fatigue life to a significant degree; 10:1 or greater is a typical reduction.

(h) Gauge Resistance and Sensitivity

The majority of gauges used for standard applications are of $100\,\Omega$ or $120\,\Omega$ resistance. Gauges are available, however, with larger resistances, e.g. 300, 500, 600 and 1000 ohm. These are used for dynamic strain investigations, and to provide larger signals which are less influenced by noise.

Most standard gauges have a gauge factor of approximately 2. For increased sensitivity, higher gauge factors are available; semi-conductor-type gauges, in particular, have gauge factors up to 70 or 80 times the standard value. These gauges are considered on page 60.

(i) Temperature Compensation

Strain gauges are sensitive to temperature. As an example, consider the case of a gauge manufactured from constantan with a temperature coefficient of resistance of -0.000012 ohm/ohm deg C raised in temperature through $3°C$.

For a $120\,\Omega$ gauge, therefore,

$$\Delta R = -0.000012 \times 120 \times 3$$
$$= -0.0043 \text{ ohm.}$$

This is almost four times the resistance change obtained previously for an accuracy corresponding to $150\,\text{lbf/in}^2$ in steel! See page 5.

When recording strain values in transient ambient conditions, therefore, it is vitally important that any temperature effect on the strain gauge should be cancelled out, leaving only the mechanical strain value required (except, of course, in the case of thermal-stress problems, where the reverse may be true).

There are two methods of obtaining temperature compensation. In the dummy gauge system (see page 5), the main requirement is that the dummy bar on which the gauge is mounted should be close to the active gauge; the two gauges thus experience the same temperature conditions. In addition, the coefficients of resistance of the two gauges should be matched exactly. It is not always possible to satisfy these requirements in practice, and alternative methods must be employed, e.g. the use of self-temperature-compensating strain gauges.

The temperature sensitivity of gauges arises from two factors:

(a) The change of the temperature coefficient of resistance (ρ)—a *nominal* constant—with temperature.

(b) The differential expansion between the gauge material and the specimen to which it is bonded. The two materials generally have different coefficients of expansion.

It is not possible to alter the coefficient of expansion of a gauge alloy but, by careful heat-treatment, it is possible to modify the coefficient of resistance within certain limits. This may be carried out in such a way that the change in resistance of the gauge due to temperature is exactly equal and opposite in sign to the change due to the differential expansion. This is possible over a restricted temperature range; self-temperature-compensated gauges can be used with complete compensation only over this range, and on the base material for which they were designed. A range of such gauges is available for use on steel, stainless steel, copper, aluminium, beryllium, etc.

Figure 13 shows a typical temperature-compensation curve for Budd metalfilm gauges.

It must be remembered that lead-wire resistance and capacitance also change with temperature, and it is therefore necessary to compensate the lead wires.

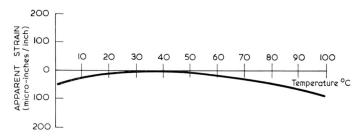

Fig. 13. Typical temperature compensation curve for Budd 101 and 104 Series Metalfilm strain gauges. Note: Curve applies to output of single gauge with change in temperature when bonded to a material having a temperature coefficient corresponding to the self temperature compensation figure of the strain gauge

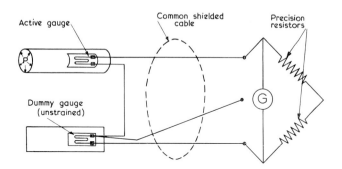

Fig. 14. Temperature-compensated three-lead wiring system

In conditions of variable temperature, the standard two-lead wiring system to strain gauges is inadequate; it introduces a variable resistance in one arm of the Wheatstone bridge, and results in a thermal signal being recorded as strain. Temperature compensation must therefore be achieved, using for example the "Siemens" three-lead wiring method as shown in Fig. 14. In this technique, two of the leads are in opposite arms of the bridge, so that their resistance changes cancel; the third lead, being in series with the power supply, does not influence the bridge balance. All leads should be of wire of the same size and length, so that they have the same resistance; they should be run in the same cable, or wound tightly together so that all experience the same temperature conditions.

In applications where long runs of wire are to be used, the gauge of the wire should be large enough to produce a lead-wire resistance which is small in comparison with the resistance of the bridge circuit.

(j) **Method of Installation and Protection**

The performance of an installation may be affected significantly by (1) the efficiency of the installation procedure, and (2) the selection of the adhesive to be used. These factors are considered in the following section.

5

STRAIN GAUGE INSTALLATION

The quality and success of a strain-gauge installation are influenced greatly by the care and precision of the installation procedure and the correct choice of adhesive[22]. The apparently mundane procedure of cementing the gauge in place is a critical step in the operation. Every precaution must be taken to obtain a clean surface prior to bonding; the surface must be free from all forms of contaminant, e.g. dust, grease and moisture.

The following is a typical surface preparation procedure used with a mild-steel specimen:

1. All paint, scale, oxide film, etc. is removed with a rotary file, scraper or rough-grade emery cloth.
2. The surface is wiped liberally with a solvent such as acetone, trichloroethylene, carbon tetrachloride, etc., to remove all grease and dirt. This process is continued until the swab or tissue remains clean when wiped across the surface.

   An area at least six times that required for the gauge itself is cleaned, to prevent dirt being drawn onto the clean portion during the installation procedure.
3. The surface is lapped, using circular motions, with silicon carbide (wet or dry) paper dipped into a 10 per cent phosphoric acid solution or commercially-available "metal conditioner".
4. The residue is wiped from the surface with a dry, clean tissue.

5. The gauge location is marked.
6. A thin layer of the acid or conditioner is applied to the surface; it is allowed to stand for a few minutes to etch the surface, and then removed with a clean tissue.
7. A thin layer of neutralizer (ammonium hydroxide) is applied to the surface, and allowed to stand for a minute before wiping off with a clean tissue.

Throughout this process it is essential that the hands should be kept clean, so that no foreign matter can be introduced onto the cleaned surface. The surface is now ready for gauge application.

The cement is applied to the surface, following carefully the procedure laid down in the manufacturer's instructions, especially where mixing of two-part components is concerned.

It is essential to keep glue-line thicknesses (i.e. distance between gauge backing and component surface) as small as possible; thick glue-lines may cause errors, especially in bending applications, and poor bonding may also result. All air must be expelled from beneath the gauge, but this must be done with care to avoid damaging the gauge element.

A neat installation may be obtained by masking the area to be used with adhesive tape. Any stray cement which spreads onto this can be removed with the tape when the gauge is in position.

The use of terminal strips with short jump-wires is recommended, as shown in Fig. 15. Heavy leads may be run from the terminal strips, and any unforeseen force on the lead wire will usually cause a break at the terminal strip instead of tearing off the gauge itself. In such cases, the terminal strip can be mounted at the same time as the gauge, using the same cement.

It is sometimes necessary to heat-cure the cement. A short heat-cure may also be beneficial when an "air-drying" cement is used.

It is preferable that the complete installation should be cycled through the load or temperature range anticipated before results are taken. In the first few cycles, load-cycling causes cold working of the gauge material, resulting in a change in resistivity and hence a zero shift. Preliminary cycling thus stabilizes the gauge and improves the accuracy of subsequent measurements.

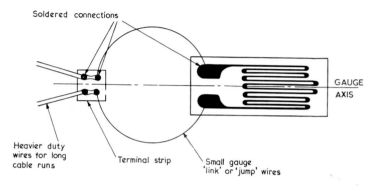

Fig. 15. Strain gauge installation showing use of a terminal strip

Testing the Installation

The following sequence of simple tests will ensure that the installation has been made without any obvious faults or damage.

1. The gauge resistance is checked before and after installation to ensure that:
   (a) it is as quoted by the manufacturers, and
   (b) no damage has occurred during the installation process.
2. The insulation resistance, i.e. resistance to earth, is checked, using a megohmeter which does not apply too high a voltage capable of causing breakdown of the gauge or adhesive. If possible, a special strain gauge megohmeter should be used. The value obtained should not be less than 1000 megohms; preferably it should be much greater, say 10,000 megohms. Values of less than

1000 megohms indicate that further curing of the adhesive layer is required.
3. The gauge is connected to a strain indicator (strain-gauge bridge) and the face of the gauge is pressed lightly with a pencil or similar object. The indicator should then record a value of strain which disappears (or very nearly so) when pressure is removed. This indicates that there is little likelihood of an air bubble or void being present under the gauge.
4. Gauge to specimen capacitance may be used to check the completeness of the adhesive cure, since the dielectric properties of adhesives are related to their state

Fig. 16. Typical strain gauge installation showing six of eight linear gauges bonded on the surface of a cylinder to record longitudinal and hoop strains. All gauges are wired back to a common terminal strip with small gauge wire. Heavier duty wire is then led from the terminal strip to suitable instrumentation. This increases the neatness of the installation and helps to prevent the gauges being pulled off or damaged in the event of the cables being inadvertently pulled

(*Crown Copyright*)

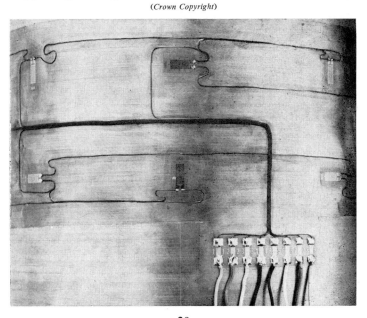

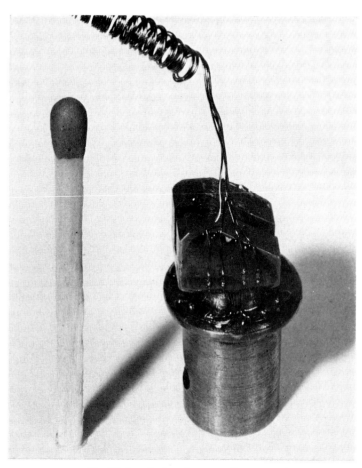

Fig. 17. Installation of very small gauges
(*Welwyn Electric Ltd.*)

of polymerization. Typical values cannot be quoted, as values vary greatly depending on gauge size and type and on the adhesive used. A value for a particular installation may be obtained by carrying out a calibration test on an installation which is known, from previous tests, to be fully cured.

Fig. 18. Strain gauge installation on a G.R.P. pressure vessel
(Hanger Engineering Ltd.)

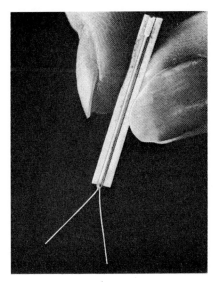

Fig. 19. Weldable strain gauge designed for high temperature and adverse environmental conditions
(Microdot Inc.)

TABLE 1
Cements Used for Strain-Gauge Measurements*

Types of Cement	One-component			Two-component	
	Soluble	Thermoplastic	Pressure-sensitive	Cold-hardening	
Base	Cellulose	Shellac	Acrylate	Polyester	Epoxy
Examples and Suppliers	PR 9241 (Philips); ST-4 (Baldwin) Duco	De Khotinsky	Eastman Kodak 910; Sicomet 85 (Sichil, Hannover) Loctite 404	PR 9244 and 46 (Philips); F-88 (Baldwin); P-2 (Kyowa)	Various hard Araldite cements (CIBA), (GY-257-F-hardener X 83/144); (103-hardener 910); Araldite hardener DP 116
Setting conditions Temp. (°C) Press. (atm) Time (h)	20–60 0·1–0·5 12–48	125 0·5 —	20 0·5 —	20 0·5–1 3–15 min.	20 0·5–1 10–50
Max. temperature for strain measurements (°C)	60		80	80 (quick-setting) (max. 180)	80
Strain-gauge type	Paper gauges	All types	Not paper	All types	All types that can stand the max. temperature
Adhesion			The cements adhere to nearly all metals in common use (not equally well to all). The adhesion to some types of plastic is poor (must be experimentally determined)		
Remarks	Moisture-sensitive; soluble in acetone and M.E.K.	Strain gauges should be able to stand a temperature of 125°C	Moisture-sensitive, do not keep well; should not be used in enclosed spaces— health risk	Does not keep very well; store at low temperatures; adhesion is sometimes critical	Advisable to store at low temperature; some hardeners are toxic

* Adapted from Potma[2].

Two-component (cont'd)		Ceramic Cements		
Warm-hardening		Hardening at High Temperature	Molten Spray	Sinter at High Tempearture
Epoxy	Phenol	Phosphate	Oxide	
Various Araldite cements (CIBA), (15-17, Ay-105), various Epikote cements (Shell) (828); Eccobond 104; EPX-150 and 400 (Baldwin) B.R. 600 (Welwyn)	PR 9243-46 (Philips) and various bakelite resins (Bakelite-Gesellschaft) B.A.P. 1 (Welwyn)	Brimor cement (Welwyn); AL-P1 Baldwin	Rokide A and C (Norton)	L-6AC (Hommel Co., (U.S.A.)
100–200	140–200	300–600	—	900
0·5–3	1–5	—	—	—
2–10	1–4	1	—	$\frac{1}{2}$
180 (max. 250)	180 (max. 250)	300–800	300–800	max. 800
All types that can stand the max. temperature	Only bakelite gauges	Only for high-temperature gauges		
Keeps well at low temperatures after mixing with hardener; some hardeners are toxic	usually delivered as *one-component* cement: keeps well	Hygroscopic; max. strain about $\frac{1}{2}\%$; poor insulation at high temperatures	Hygroscopic; max. strain 2%; poor insulation at high temperatures	Max. strain about $\frac{1}{2}\%$

6

GAUGE PROTECTION

When the gauge has been installed, it must be protected adequately from damage or deterioration. The nature of the protection will depend upon the environment and conditions under which the gauge is to be used[17]. The following factors are commonly of importance in this respect:

(a) water; including the effects of moisture and humidity, or contact with liquid water or brine;
(b) mechanical damage;
(c) other hostile environments; contact with oil, petrol, acid, etc.

(a) **Water**

Strain gauge installations may be affected by water in many ways. Water may attack the insulation, reducing its resistance; the effect is comparable with that of placing a shunt resistance across the gauge. The strength and rigidity of the adhesive bond may be reduced, resulting in poor strain transference from specimen to gauge element, a change of gauge factor, and creep and hysteresis problems.

In addition, the presence of moisture in some adhesives can make them swell; they will subsequently contract if the moisture is driven off at a higher temperature. This causes strain on the gauge element, which will be recorded as part of the measured "mechanical" strain.

In the presence of moisture, electrolysis may take place when current is passed through the gauge, causing erosion of the gauge filament. This increases the resistance of the gauge, and introduces weaknesses which result in early fatigue failure.

There are many methods available for protection against moisture, the choice of method depending on the severity of conditions. For normal laboratory applications there are many commercially-available gauge coatings; these provide adequate protection, provided that the jump leads and terminal strips (if any), in addition to the gauge itself, are coated.

For applications which must withstand more severe conditions, such as prolonged contact with seawater, protection must be increased accordingly. Thicker coatings may be necessary, and care must be taken to ensure that the resulting stiffness does not adversely affect readings. (The life of a typical water-proofing coating is roughly proportional to its thickness, which should be not less than $\frac{1}{16}$ in to ensure efficient protection over long periods. Several thin coats provide more effective water-proofing than a single thick coat.)

A typical water-proofing coating has an initial coating of wax, silicone rubber or Bostik compound. This is covered with a special metal shim dome or cover, and the whole unit is then covered with a second coating of rubber compound.

Particular care should be exercised at the coating/lead-wire seal, as this is usually where the protection breaks down. P.V.C.-coated wires may cause difficulty, as adhesion tends to be poor; rubber-coated wires are often preferred for this reason.

(b) **Mechanical Damage**

A degree of protection from mechanical damage can be achieved by coating the installation lightly with silicone rubber. More adequate protection may be achieved by using methods similar to those described in (a) above. The installation is first coated with a rubber compound or an epoxy adhesive, and the coating is allowed to cure. The metal dome, or a piece of fibreglass cloth impregnated with adhesive, is then placed over the coating, and a further heavy coat of rubber or adhesive applied.

(c) Other Hostile Environments

Proprietary coatings are available which provide adequate protection against most of the materials and environmental conditions commonly encountered. The manufacturers of coatings will always advise the user about the degree of protection afforded in specific circumstances, or against particular chemicals.

7

EFFECT OF STRAIN GAUGE EXCITATION CURRENT

Variations in ambient temperature can exert a significant influence on the output from strain gauges, particularly those which are not self-temperature-compensating. The gauge current, in its turn, significantly affects the temperature of the gauge, so much so that the temperature fluctuations experienced under normal laboratory conditions are too small to be of significance.

When it is connected in the Wheatstone bridge circuit, the gauge experiences a current I; this produces heat, since power must be dissipated according to the relationship

$$\text{power} = I^2 R$$

where R is the gauge resistance.

As the heat is dissipated, principally by conduction to the component surface, the temperature of the gauge rises above the ambient temperature. Gauge performance will then be affected in the following ways:

1. When using self-temperature-compensating gauges, some compensation is lost as the gauge-element temperature and the specimen temperature may differ significantly. Data published by the manufacturers of gauges relate, of necessity, to low excitation levels.
2. Hysteresis and creep effects are magnified, since they are dependent on backing and adhesive temperatures.
3. Gauges can experience a considerable "zero shift", i.e zero-load stability is influenced strongly by high excitation rates.

4. Any voids or bubbles in the adhesive, or discontinuities in the backing of the gauge, can cause "hot-spots" and localized areas then operate at much higher temperatures than the rest of the gauge. This effect produces creep and instability problems, and only high-quality gauges, expertly bonded, should be considered for high-excitation applications.

The heating effect is dependent on a number of factors, including gauge size, type and configuration, adhesive type and thickness, waterproofing or protection material, specimen material and size, etc. The following rules serve as a useful guide.

(a) With respect to their efficiency in dissipating heat, gauges may be placed generally in the following order:
   1. Foil gauges,
   2. Flat wire gauges,
   3. Round-wound type wire gauges.

(b) When gauges are used on poor conductors, such as plastics, ceramics, concrete, etc., gauge currents must be considered extremely carefully. Poor heat conduction can cause large errors, even with normal excitation levels.

(c) Typical gauge currents of, say, 25 ma are relatively safe on materials of high conductivity. In cases where higher sensitivity is required and higher currents are necessary, the gauge stability should be checked as follows:

The bridge excitation under zero load is increased gradually until a definite zero instability is observed. It is then reduced until the zero reading is once more stable, and near the low-excitation zero reading. This value is then the highest excitation which can be used without loss of performance. In thermal applications, this test should be carried out at the maximum operating temperature rather than at room temperature.

(d) Adhesive glue-lines should be kept as thin as possible to aid heat dissipation. Typical values of glue-line thickness for smooth surfaces range from 0·0001–0·0005 in.

(e) At high excitation levels where zero shift is sometimes accepted, within limits, time must be allowed for gauges to become stable as temperature equilibrium is reached.

8

HIGH-TEMPERATURE MEASUREMENTS

Three main systems are used in bonding gauges for high-temperature applications.

(a) Ceramic Cements.
(b) Rokide Process.
(c) Welding.

(a) **Ceramic Cements**[18,13]

Brimor cement and Nichrome V gauges are used in the ceramic cement process. This is a reasonably simple system which can provide an installation with good fatigue life.

A typical procedure would be as follows:
1. The surface is sandblasted.
2. One to three precoats of Brimor are applied, each being cured at 350°C.
3. The gauge is applied.
4. Two overcoats of Brimor are applied, each being cured at 350°C.
5. The leads are attached and secured with two to four coats of cement, depending on the severity of the application, each being cured at 350°C.

(b) **Rokide Process**[14,15]

In the Rokide process, a ceramic rod is passed through an oxy-acetylene flame; molten particles of ceramic, atomized by and carried in a stream of compressed air, are deposited on the suitably-prepared surface of the component. They solidify to form a dense and homogeneous layer which is highly resistant to wear, to mechanical or thermal shock, and to the effects of hostile environments (including most chemicals). The electrical insulation is excellent.

This system was developed initially to protect corrosive materials, but it has now been adopted widely for the mounting and protection of strain gauges used in high-temperature work.

The bonding of Rokide is effected principally by mechanical keying, which depends on the size of the particles and their surface contact. Rokide is normally used in conjunction with frame-mounted gauges which leave portions of the element grid exposed.

A typical procedure would be as follows:

1. The component should be cleaned thoroughly, and grit-blasted.
   Rolls Royce[13] claim that better adhesion is obtained by using either (a) Bauxelite blast followed by flash coat of Metco 404, or (b) Vapour (400 Grade) blast followed by flash coat of Metco 404.
   The Bauxelite blast is not suitable for thin sections because of the heavy surface damage caused.
2. If Metco is not used as suggested above, a pre-coat of Nichrome should be applied immediately after blasting. If temperatures are not likely to exceed 400°C, molybdenum may be used. In each case, the thickness of pre-coated material should not exceed 0·002 in.
3. An insulation layer of Rokide, 0·004 to 0·005 in thick, is sprayed on.
4. The gauge is placed on the surface and clamped in position, e.g. with the aid of magnets, clamps, bulldog clips, etc.
5. The surface of the component is warmed to the temperature at which the insulation coat was deposited.
6. The gauge is secured in position by spraying Rokide from a distance of approximately 15in onto the exposed parts of the grid.

7. The clamping devices are removed. The surface is warmed again before finally coating the whole assembly, gauge and leads, to complete the installation.

A modified procedure may be used for mounting other types of gauges with Rokide. Details can be obtained from the manufacturer.

A disadvantage of this technique is that spraying should always be carried out normal to the surface. This is not always possible in certain positions, and a hybrid system has been suggested by Bridge and Falcus; a base coat is sprayed onto the component, the gauge is fixed with Brimor and the leads are anchored with Rokide.

Both the Rokide and Brimor systems have been used, with suitable gauge selection, up to 850°C static and 1000°C dynamic measurements.

(c) **Welding**

A folded-filament gauge is available from Microdot, U.S.A.; it can be welded directly to the surface and is suitable for temperatures up to about 400°C. It is available in temperature-compensated form. According to its manufacturers, the gauge is "expensive but features rapid installation, high resistance to ground, dependable operation, high allowable strain levels and low apparent strain versus temperature".

Other gauges are available, e.g. from B.L.H., which have been pre-mounted on a metal shim. This, in turn, can be welded to the surface under test.

9

LONG-TERM TESTING[25]

Problems associated with long-term testing using strain gauge installations are essentially:

(a) Instrumentation zero drift,
(b) Lead wire drift,
(c) Strain gauge drift.

If meaningful data are to be obtained over a period of months or even years, it is essential that every precaution be taken to ensure the correct choice of strain gauge type, perfect installation and curing procedure, and thorough monitoring of the above sources of error.

(a) Instrumentation Zero Drift

The validity and stability of an instrument true zero is obtained as follows:

A full Wheatstone bridge circuit of temperature-compensated foil strain gauges is prepared which can be kept in a laboratory. This is connected to the instrument in question, and the zero reading recorded. The input leads at the instrument are reversed, and the reading is again recorded. The algebraic difference of the two readings divided by two is then the true instrument zero. This zero should be monitored before each measurement is taken, and the subsequent readings related to it.

A low voltage excitation should be used for the gauges; this reduces gauge-current heating to a minimum, and allows sufficient time prior to readings for the gauge current to heat the gauges until thermal equilibrium is obtained.

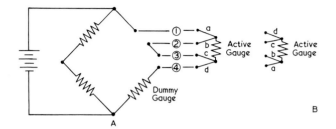

Fig. 20. Wiring system to produce lead wire drift elimination

(b) **Lead Wire Drift**

This error occurs as a result of corrosion and strain on the leads. It can be cancelled completely, however, by using a single active gauge and the four-lead wiring system shown in Figs. 20(A) and 20(B).

Two measurements are made, one with the circuit of Fig. 20(A) and the other with that of Fig. 20(B). The average of these readings (minus the true instrument zero) is the true strain reading.

(c) **Strain Gauge Drift**

This drift cannot be monitored or cancelled, and it represents the most serious source of possible error. The success of any installation relies largely on the measure of gauge stability that can be ensured.

Foil gauges of the advance-temperature-compensated type are to be preferred; foil gauges, particularly of large gauge area, place less stress on the adhesive used and thus limit creep to a minimum. Gauge lengths should be as long as possible to benefit from the better heat-dissipation properties of longer gauges.

Adhesives used are usually of the Bakelite or epoxy type. More effective polymerization and lower creep effects will be obtained, especially with an epoxy adhesive system, by using a post cure to, say, 180°F for one week. Particular care should be taken with waterproofing, following procedures similar to those described on page 34. Where possible, the gauges should be immersed in water for three to four days prior to test. If the insulation resistance is then in excess of 500 megohm the installation has every chance of success.

10

DYNAMIC MEASUREMENTS FROM ROTATING COMPONENTS

When strain measurement is to be carried out on rotating components such as shafts, wheels, turbine blades, propellers, etc., two methods are commonly used: (a) slip rings and (b) telemetry.

Slip Rings

A full discussion of the use of conventional brush and slip ring assemblies, or more advanced types of mercury slip ring, is beyond the scope of this text. Information may be obtained from references 9, 10 and 11.

Table 2 lists typical materials used for brushes and slip rings to produce minimum noise. These are included for the guidance of the reader who contemplates manufacturing his own slip ring assemblies. It should be stressed, however, that the problem of noise is not easily overcome, except by making use of the second technique, i.e. telemetry.

"Stick-on" slip rings are now available to facilitate the taking of strain measurements from rotating shafts. It is not suggested that this type of assembly will provide the performance attainable with precision-manufactured components, but such assemblies have operated satisfactorily for the equivalent of 5000 miles per slip ring, without wear and with negligble noise level for speeds up to 25 ft/sec.

Telemetry[11]

A second measurement technique, based on radio transmission from the rotating object, was developed at the

Central Electricity Research Laboratories at Leatherhead, England.

Conventional brush and slip ring systems present other problems, in addition to wear and noise, when operated at high speeds in difficult environments; in steam turbines, for example, the environment will include high temperatures, centrifugal fields of perhaps 7000 g, and a wet steam atmosphere.

A telemetry system has been developed which operates successfully under these conditions. It does not involve physical contact between the rotating component and the stationary recording instrument. The transducer (strain gauge) signals are amplified and transmitted at radio frequencies using frequency modulation.

Information on this technique can be obtained from reference 11. Its success may be judged by the fact that U.K. turbine manufacturers have adopted the system for studying L.P. turbine blade behaviour up to 500 MW capacity.

TABLE 2
Materials for Slip Rings and Brushes

Brushes	Typical Pressure		Slip Rings
	lbf/in^2	KN/m^2	
Carbon	40 to 80	300 to 600	Silver
Carbon	70 to 140	500 to 1000	Brass
Carbon (soft)	80 to 160	600 to 1200	Monel
Carbon (hard)	100 to 210	700 to 1500	Inconel
Silver Graphite (40% C)	40 to 80	300 to 600	Brass
Silver Graphite (40% C)	40 to 80	300 to 600	Monel*
Silver Graphite (40% C)	40 to 80	300 to 600	Inconel
Copper Graphite (50% C)	70 to 140	500 to 1000	Stainless Steel

* In most applications the Silver-Graphite/Monel combination is to be preferred.

Extensions to the system have been used, for example, for measurements on concrete dams, bridges, etc., where very long leads would otherwise be required, for torque and temperature measurements on shafts and for alternator rotor temperature measurements.

11

BASIC MEASUREMENT SYSTEMS

Quarter-Bridge System

Figure 21 shows a Wheatstone bridge circuit including one active gauge Rg, the remainder of the arms being completed with standard resistors. This is called a quarter bridge.

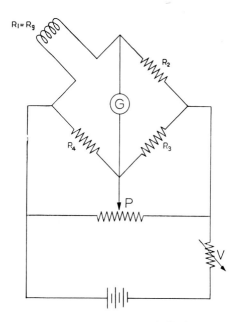

Fig. 21. 'Quarter-bridge' circuit

Included in the circuit is a potentiometer *P* which can be used to balance the galvanometer to zero with no load conditions; a variable resistance *V* provides an adjustment of bridge excitation.

This circuit, using standard gauges, is not temperature compensated, and it should be used with caution. It is, however, useful in temperature-controlled laboratory conditions for the measurement of dynamic strains, or for measurements using a single temperature-compensated gauge.

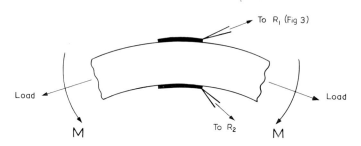

Fig. 22. Determination of bending strains independent of end loads

Half-Bridge System

In the half-bridge system, two resistances of the Wheatstone bridge network are precision resistors (supplied in the instrument itself); the other two are either one dummy and one active gauge to provide temperature compensation, or two active gauges.

Where two active gauges are used, there are two alternative systems:

1. *Gauges in adjacent arms of the bridge*

This system is particularly useful, for example, when it is necessary to separate bending strains from plane strains.

Consider a beam subjected to end load and bending, with

strain gauges mounted on the top and bottom faces as shown in Fig. 22.

If gauge 1 is connected as R_1 and gauge 2 as R_2, the similar signals due to end load experienced by both gauges will be cancelled, as they are on opposite "sides" of the bridge. The bending effects, however, being opposite in sign, will both produce deflection of the galvanometer; the resultant reading will be the sum of the bending strains on the two gauges. Torsional effects are also cancelled.

2. *Gauges in opposite arms of the bridge*

The effects of gauges in opposite arms of the bridge, say R_1 and R_3, will reinforce each other. In the example described above, the strain due to direct load on each gauge will be added together, whereas the bending strains, being opposite in sign, will cancel each other. Thus, a signal twice that of a single gauge is again achieved, this time providing information of the direct-load effects, independent of bending.

In each case, temperature compensation is achieved using self-temperature-compensated gauges and the three-wire-lead system.

Full-Bridge System

In the full-bridge system, all four arms of the Wheatstone bridge contain active gauges, providing maximum sensitivity.

Consider, for example, the case of a bar in tension. The maximum signal for a given load is obtained by using two gauges along the axis of the bar (one on the front, the other on the back of the bar), and two gauges at right angles to the axis to record the Poisson's ratio strains, which will be compressive. Since Poisson's ratio for most engineering materials is approximately 0·3, the two latter gauges will record strain values 0·3 times the magnitude of, and opposite in sign to, those of the gauges along the axis. Connecting the circuit as shown in Fig. 23 produces a summation of the strain signals from all four gauges, i.e. a magnification of 2·6 times.

Similarly, two gauges on each of the top and bottom faces

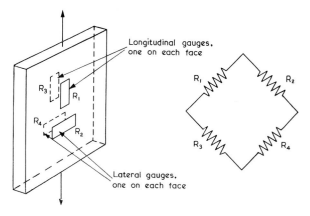

Fig. 23. 'Full bridge' circuit arranged to eliminate any bending strains present due to eccentricity of load and to achieve a sensitivity approximately 2.6 times that of a single active gauge

of a beam in bending would produce four times the output of a single gauge. In addition, the gauges could be arranged to cancel all direct load and torsional effects.

Temperature compensation is best achieved by using self-compensating gauges, and by wiring all four gauges with identical leads to a terminal block placed nearby. Four identical wires from the terminal block to the amplifier, run in a common shielded cable, will provide lead-wire temperature compensation.

The full bridge circuit is also useful in that it will yield four times magnification from gauges mounted on a shaft to record torsion independent of axial load and bending.

The increase in sensitivity achieved with this circuit makes it very attractive in strain-gauged load cell and transducer design.

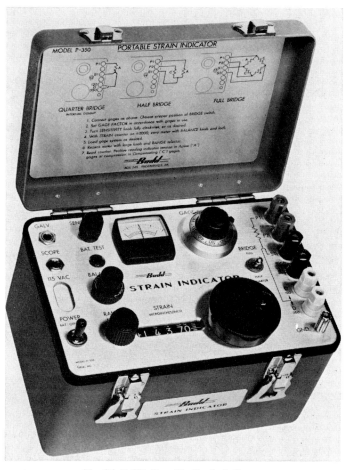

Fig. 24. P.350 Portable Strain Indicator
(*Automation Industries (U K.) Ltd.*)

12

D.C. AND A.C. SYSTEMS

Both d.c. and a.c. measurement systems are used in strain gauge work, the relative advantages and disadvantages of the two systems being indicated in Table 3.

An advantage of the a.c. carrier frequency system is that all unwanted signals, such as noise, are eliminated, and a stable signal of gauge output is produced. A typical carrier frequency system is shown in Fig. 25.

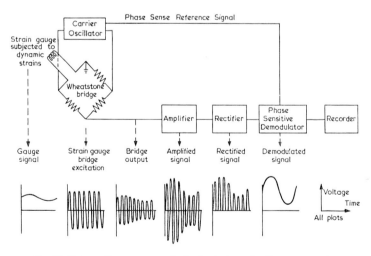

Fig. 25. Schematic arrangement of a typical carrier frequency system

TABLE 3

Advantages and Disadvantages of D.C. and A.C. Measuring Systems

D.C.	A.C.
Notably subject to unwanted signals from: 1. Noise. 2. Contact potentials ("battery effect") 3. Thermoelectric voltages (thermocouple effect at joint of two dissimilar metals).	Free from unwanted signals listed opposite.
	Very high gain possible.
Relatively inexpensive.	Relatively expensive.
Can be portable.	
Unsuitable for static and low frequency dynamic work.	Suitable for static and low frequency dynamic work.
Suitable for high frequency measurements.	Unsuitable for high frequency measurements.
Useful when long leads are necessary, since this eliminates capacitance effects.	
If a D.C. amplifier is required in circuit the unit becomes complicated and expensive.	A.C. amplifiers simpler in construction and more reliable.
Zero adjustment simple; no phase balancing required.	Phase balancing required in addition to resistance balancing, since the small self inductance and capacitance of each resistance will have an effect on the balance of the a.c. system.

(continued over)

TABLE 3 (cont'd)

D.C.	A.C.
No interference from inductance and capacitance, i.e. unscreened cables can be used.	Inductance and capacitance problems, as above. Screened cables required.
Bridges may be combined for multi-point measurements.	Addition of bridges difficult, as there will be interaction between the bridges, and balancing of one will affect the balancing of others.
	Mains supply can be used.

An oscillator energizes the normal Wheatstone bridge arrangement, the gauges being manually-balanced to take account of any slight resistance and capacitance differences in the four arms, so yielding zero load balance. When the gauge is strained, the bridge is unbalanced and the carrier signal produced is modulated by the applied strain input. This modulated carrier signal is then amplified, rectified and demodulated as shown; the final noise-free signal is recorded on a direct-writing recorder, providing a permanent record of the strain signal. Very high gains are possible with this type of measurement system.

Strain gauge bridges may be energized, using either sine-wave or square-wave excitation; the relative merits and disadvantages are indicated in Table 4.

TABLE 4

Relative Merits of Sine Wave and Square Wave Excitation

	Sine Wave	Square Wave
Capacitance effects	Shifts phase and reduces final signal. Capacitive balancing required.	Alters shape of square wave. Less severe effect, as it can still be demodulated. No capacitive balancing required.
Dynamic recording	Better response at high frequency (100 Hz or more). Limited only by the frequency of the input. Approximately suitable for $\frac{1}{10}$–$\frac{1}{5}$ the frequency of the carrier signal, e.g. 1000 Hz carrier suitable for 200 Hz strain gauge signal. Capacitance troubles increase with frequency; practical carrier limit of the order 1–5 kHz.	Reduced signal as frequency increases. [graph: % Signal vs Log Frequency (Hz), showing curve dropping from 100 at low frequency through 80 around 50 Hz and 40 near 100 Hz to 0 by 200 Hz] Calibration is possible to take account of the first 30 per cent drop.

13

RECORDING SYSTEMS

Table 5 lists the methods of recording which are available for the complete range of strain signals normally encountered.

TABLE 5

Recording Systems

Frequency of Signal	System Preferred	Recorder Type Applicable
Static	A.C. (preferably) or D.C.	Null balance indicator or data logger
Quasi-static		Null balance and servo system to motor the balance slidewire.
Up to 100 Hz	A.C. or D.C.	Pen and ink recorder.
Lower 100s Hz		Electrically-sensitive paper with voltage applied to stylus, OR Hot pen and wax paper.
Upper 100s Hz	D.C.	High voltage stylus and heat sensitive paper.
1000s Hz	D.C.	Ultra-violet recorder.
Mega-hertz	D.C.	Oscilloscope

14

OTHER TYPES OF STRAIN GAUGE

1. Mechanical Gauge

Mechanical strain gauges, i.e. extensometers, are used effectively in many applications, including calibration work in the laboratory, standard tensile and compression testing, and measurements on site using, for example, Demec gauges.

These gauges, using mechanical lever or optical lever principles, achieve a wide range of sensitivity. The accuracy attainable can be comparable with that of the electric resistance strain gauge. Mechanical gauges are, however, bulkier and less flexible in application than electrical strain gauges.

2. Pneumatic Gauge

A pneumatic gauge, working on a sonic orifice principle, has been developed at R.A.E. Farnborough, England. Changes in strain cause changes in pressure to be recorded on a gauge upstream of the orifice, and this can be calibrated to give strain. The unit was developed primarily to overcome the limitations of conventional resistance-wire gauges, especially at high temperatures.

3. Vibrating-Wire Gauge (Acoustic Gauge)[30]

The operation of the vibrating-wire or acoustic gauge is governed by Mersenne's Law, which relates frequency of vibration of a tensioned wire with the applied strain as follows:

$$f = \frac{1}{2L}\sqrt{\frac{T}{m}}$$

where f = frequency of vibration.
L = effective length of wire.
T = tension in the wire.
m = mass per unit length of the wire.

This can be re-written in the form

$$f = K\sqrt{\varepsilon}$$

where $$K = \frac{1}{2L}\sqrt{\frac{AE}{m}}$$

ε = strain
E = Young's modulus
and A = cross-sectional area of the wire.

Thus, the frequency f will vary with increasing strain; variations can be calibrated to give strain directly.

The gauge consists of a length of wire stretched between two posts. The wire is plucked, the frequency of the resulting vibrations being recorded automatically on suitable instrumentation. The gauges are usually enclosed in briquettes, and are built into civil engineering structures such as roads, dams, bridges, etc. They are particularly useful for long-term applications. Strains can be monitored throughout the service history of the structure.

4. Piezo-Resistive (Semi-Conductor) Gauge[12]

The heart of this gauge is a filament consisting of a single crystal of silicon. Suitable leads are attached, and the gauge is available with or without backing. The action of the gauge depends upon the piezo-resistive effect in silicon, which is much more sensitive to dimensional changes than is the electrical resistance. The sensitivity of a piezo-resistive gauge is up to 70 times greater than that of conventional gauges,

Fig. 26. Small, inexpensive, battery-operated portable strain gauge bridge, particularly useful for field work
(*Strainstall Ltd.*)

and both positive or negative resistance changes are available from a single strain type.

Gauge factors are available from -140 to $+140$, resistance values up to 2000 ohms and gauge lengths from 1 inch to 0·02 inch.

Non-linearity is a problem with piezo-electric gauges, and relatively sophisticated instrumentation is required.

Semi-conductor gauges are used extensively for ultra-low-level static strain measurements and for transducer applications. In most general strain work, exceptionally high sensitivity is not necessary, and the more robust and economical foil strain gauge is perfectly adequate.

5. Inductance Gauge

There are many types of inductive strain-gauge system, the differential transformer system[29] being the most popular.

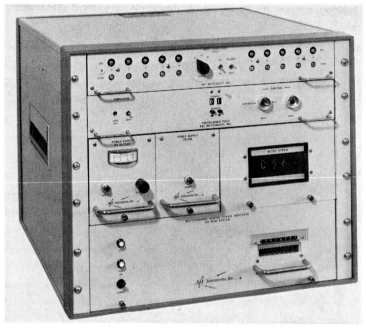

Fig. 27. B & F Instruments Inc. Multi-channel Strain Indicator
(*Welwyn Electric Ltd.*)

A typical linear differential transformer arrangement is shown in Fig. 28. As the core moves within the coils, the inductance coupling between the primary and secondary windings is varied; the voltage signal produced can be related to the displacement of the core. The system can also be employed to measure pressure, displacement, temperature, force and acceleration.

6. Capacitance Gauge[28]

The capacitance gauge consists essentially of two parallel plates as shown in Fig. 29. The capacitance between the plates is determined by the relationship

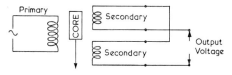

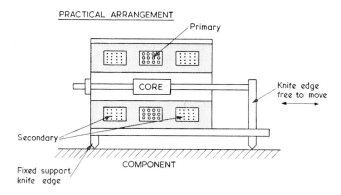

Fig. 28. Linear differential transformer strain gauge

$$C = 0\cdot 225\frac{kA}{h}$$

where C = capacitance

A = effective cross-sectional area of plates

h = distance between plates

k = dielectric constant of the medium between the plates.

This system can be arranged to measure strain or displacement by

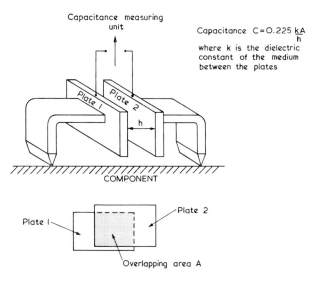

Fig. 29. Basic arrangement of a capacitance strain gauge

(a) changing the gap distance H, or
(b) moving the plates transversely with respect to each other, thereby changing A, the effective area between the plates, or
(c) by moving some unit with a k value higher than air between the plates.

Special circuits are used to record the small capacitance changes.

The relatively-large size of this type of gauge is a disadvantage.

15

ADVANTAGES OF ELECTRICAL-RESISTANCE STRAIN GAUGE TECHNIQUE IN EXPERIMENTAL STRESS ANALYSIS

1. It is relatively simple and quick.
2. Readings are direct.
3. Accuracy is extremely high.
4. Sensitivity is high.
5. Interpretation of results is minimal.
6. It is repeatable.
7. It is non-destructive.
8. It is applicable equally to static, dynamic and transient conditions.
9. Its application is not limited by component size.
10. Results can be obtained under service conditions, using service loads.
11. Numerous types, size and form of gauge exist, covering practically all applications, including the monitoring of fatigue damage.
12. The method is particularly powerful when used in conjunction with brittle lacquers. ([31])
13. There is no limit to the weight and size of gauge.
14. Service environments can be catered for.
15. Remote readings are possible in relatively inaccessible regions, or in regions which may become dangerous under load.
16. Multi-point measurements can be quickly and regularly monitored using data-logging facilities. Tape outputs from such data loggers are readily analysed using modern computer techniques.

17. Temperature effects can be largely eliminated, and measurements are possible up to 1000°C.
18. Stability and repeatability are high; strain gauges are the sensing elements of most modern load cells and transducers.
19. Bending, torsion and direct stress can be separated using various wiring systems.
20. Sensitivity may be varied by modification of the wiring system used.
21. Long-term measurements over periods of years are possible, provided adequate care is taken with the installation.
22. Extensive and valuable information can be obtained with relatively inexpensive equipment.
23. The state of the art is such that there is practically no limitation on the application of the technique.
24. Commercial development and experience since 1939 have been extensive. Advice and information are available over a wide range of applications.

BIBLIOGRAPHY

1. Perry and Lissner, *Strain Gauge Primer*, McGraw-Hill: New York.
2. Potma, *Strain Gauges*, Iliffe: London.
3. Hollister, *Experimental Stress Analysis*, Cambridge University Press: London.
4. E. J. Hearn, *Problems on Strength of Materials—Complex Stress and Strain*, Longmans: London.
5. De Forrest and Leaderman, *The Development of Electrical Strain Gauges*, N.A.C.A. Tech. Note 744, 1940.
6. Nielsen, "Strain Gauges", *Electronics*, Dec. 1943.
7. G. F. Chalmers, "Some thoughts on the making of reliable strain gauge measurements", *Strain*, **3**, No. 2.
8. Dohrenwend and Mehaffey, "Measurement of dynamic strain", *J. Appl. Mech.*, **10**, A.85 (1943).
9. Curtiss, "Stresses in rotating shafts", *Electronics*, July 1945.
10. Murray and Stein, *Strain Gauge Techniques*, M.I.T.: Cambridge, Mass. (1956).
11. *Rotating Contactless Telemetry*, Information Leaflet, C.E.R.L. Laboratories: Leatherhead.
12. *Silicon Piezoresistive Strain Gauge Element*, Paper presented at B.S.S.M. Meeting, Manchester, March 1965, Ferranti Ltd.: Manchester.
13. P. Bridge, *Strain Gauge Practice at Rolls-Royce (Derby) as Applied to High Temperature Installation*, Paper presented at B.S.S.M. Annual Conference, 1968, Coventry.
14. *Rokide*, Norton Abrasives Ltd.: Welwyn Garden City, Herts.
15. *Suggestions for the Mounting of Strain Gauges using the Norton Rokide Process*, Norton Abrasives Ltd.: Welwyn Garden City.
16. L. G. Phillips, "Self adhesive strain gauges", *R.A.E. Tech. Report*, **67025** (Jan., 1967).
17. "Strain measurement in hostile environment", *Applied Mechanics Rev.* (Jan., 1965).
18. "An investigation into the electrical leakage and limiting strains of ceramic cements suitable for mounting high temperature strain gauges", *C.E.G.B. Report, No. RD/B/N792*, **April, 1967**.
19. "Elimination of lead wire errors in the SR4 temperature compensated strain gauge", *Testing Topics*, **6** (1951).
20. B. Hague, *Alternating Current Bridge Methods*, Pitman: New York, 1938.

21. K. E. Kern, "The stress gauge", *Proc. S.E.S.A.*, **IV**, No. 1, 124–129 (1946).
22. J. Hurd, "Adhesives guide", *Brit. Sci. Instr. Res. Assoc. Report*, **M39** (1959).
23. "Some untold chapters in the story of the metalfilm strain gauge", *Absts. 734, Strain Gauge Readings*, **III**, No. 5 (1960).
24. P. K. Stein, *Advanced Strain Gauge Techniques*, Stein Engineering Services: Phoenix, Arizona, 1962.
25. "Use of electric resistivity strain gauges over long periods of time", *Proc. S.E.S.A.*, **III**, No. 2, 47–52 (1946).
26. "Transverse sensitivity of bonded strain gauges", *Wu.C.T. Expt'l Mech.*, **2,** 11 (1962).
27. "Strain gauge measurements in regions of high stress gradient", *Expt'l Mech.*, **1,** 6 (1961).
28. "Measurements of displacement and strain by capacity methods", *Proc. I. Mech. E.*, **152** (1945).
29. H. Schaevitz, "The linear variable differential transformer", *Proc. S.E.S.A.*, **IV,** 2 (1947).
30. R. S. Jerret, "The acoustic strain gauge", *J. Sci. Inst.*, **22,** 2 (1945).
31. E. J. Hearn, *Brittle Lacquers*, Merrow Publishing Co. Ltd., Watford England (1971).

APPENDIX

FIRMS MANUFACTURING STRAIN GAUGES, STRAIN GAUGE INSTRUMENTATION, ANCILLARY EQUIPMENT AND ASSOCIATED RECORDING EQUIPMENT

There are many firms and organizations concerned with strain gauge techniques in almost every country in the world. The following list represents only a cross-section of firms interested in strain gauge and associated recording equipment. It is not intended to be comprehensive.

Ad Auriema Ltd., Inspection House, 125 Gunnersbury Lane, Acton, W.3.
Addo Ltd., Addo House, 85 Great North Road, Hartfield, Herts.
Advance Instruments, Roebuck Road, Hainault.
A.E. Electronics, Glacier House, Ealing Road, Alperton, Wembley, Middsx.
A.E.I., Process Control Department, P.O. Box No. 1, Harlow, Essex.
Aim Electronics Ltd., 8 North Street, Cambridge.
Airmec Instruments Ltd., Coronation Road, High Wycombe, Bucks.
Allen West Automation Ltd., Lewes Road, Brighton 7, Sussex.
Analog Devices Ltd., 38–40 Fife Road, Kingston-upon-Thames, Surrey.
Automation Industries (U.K.) Ltd., Albert St., Fleet, Aldershot, Hants.
Automatic Systems Laboratories, Grovebury Road, Leighton Buzzard, Beds.
Aveley Electric Ltd., South Ockenden, Essex.
Avo Ltd., 92 Vauxhall Bridge Road, London, S.W.1.
Baldwin-Lima-Hamilton, c/o Mr. L. Brooke Edwards, 25 Palmeira Drive, Hove 3, Sussex.
B. & K. Instruments Ltd., 59 Union Street, London, S.E.1.
B. & K. Laboratories Ltd., 4 Tilney Street, Park Lane, London, W.1.
Bell & Howell Ltd., Lennox Road, Basingstoke, Hants.
Belling & Lee Ltd., Gt. Cambridge Road, Enfield, Middsx.
Bendix Electronics Ltd., High Church Street, New Basford, Nottingham.
Benson-Lehner Ltd., West Quay Road, Southampton.
Bourns (Trimpot) Ltd., Suite 31, Hodford House, 17–27 High Street, Hounslow, Middsx.
G. & E. Bradley Ltd., Electrical House, Neasden, London, N.W.10.
British Aircraft Corporation (Operating) Ltd., Six Hills Way, Stevenage, Herts.
Bryans Ltd., 1 and 15 Willow Lane, Mitcham, Surrey.
C.I.B.A. (A.R.L.) Ltd., Duxford, Cambridge.

C.N.S. Instruments, 61 Holmes Road, London, N.W.5.
Comark Electronics, Gloucester Road, Littlehampton.
Computer Controls Ltd., Division of Kynmore Engineering Co. Ltd., 19 Buckingham Street, London, W.C.2.
Coutant Electronics Ltd., 3 Trafford Road, Richfield Estate, Reading, Berks.
Coutant Transducers, Milford Road, Reading, Berks.
Dale Electronics Ltd., 109 Jermyn Street, London, S.W.1.
Dana Laboratories U.K., Bilton Way, Dallow Road, Luton, Beds.
Data Acquisition Ltd., 5 Ashley Drive, Bramhall, Stockport, Ches.
S. Davall & Sons Ltd., 1 Wadsworth Road, Greenford, Middsx.
Davey United Instruments, Darnall Works, Sheffield, S.9 4EX.
F. T. Davis (Kings Langley) Ltd., Kings Langley, Herts.
Daystrom Ltd., Gloucester.
Deakin Instrumentation Ltd., 66 High Street, Walton-on-Thames, Surrey.
Deakin Phillips Electronics, T. Deakin Esq., Tilly's Lane, High Street, Staines, Middsx.
Digital Engineering Co. Ltd., 22 Lombard Road, London, S.W.11.
Disa Elektronik A/s., 116 College Road, Harrow, Middsx.
Dymar Electronics, Rembrandt House, Wippen Dell Road, Watford, Herts.
Dynamco, Dynamco House, Hanworth Lane, Chertsey, Surrey.
Elcomatic Ltd., 101 Bath Street, Glasgow, C.2.
Electro Mechanisms Ltd., 218/221 Bedford Avenue, Trading Estate, Slough, Bucks.
Electronic Associates Ltd., Victoria Road, Burgess Hill, Sussex.
Elliott Brothers (London) Ltd., Servo Components Division, Century Works, Conington Road, London, S.E.13.
Elliott Bros. Ltd., Airport Works, Rochester, Kent.
Elliot Mechanical Automation, Industrial Weighing Div., Century Works, Lewisham, London, S.E.13.
Endevco Corporation, U.K. Branch, The Sycamores, Kneesworth Street, Royston, Herts.
English Electric Co. Ltd., English Electric House, Strand, London, W.C.2.
Environmental Equipments Ltd., Denton Road, Wokingham, Berks.
Epsylon Industries, Faggs Road, Feltham, Middsx.
Ether Ltd., Caxton Way, Stevenage, Herts. (Semi-conductor gauges)
Fairey Surveys Ltd., Research and Eng. Div. Reform Road, Maidenhead, Berks.
Farnell Instruments Ltd., Sandbeck Way, Wetherby, Yorks.
Fenlow Electronics Ltd., Springfield Lane, Weybridge, Surrey.
Foster Instruments Co. Ltd., Pixmore Avenue, Letchworth, Herts.
Fylde Electronics, 6–16 Oakham Court, Preston, PR1 3XT.
GEC/AEI Automation Ltd., New Parks, Leicester.
General Test Instruments Ltd., Gloucester Trading Estate, Hucclecote, Glos.
Guest Electronics Ltd., Nicholas House, Brigstock Road, Thornton Heath, Surrey.
Gulton Industries, Regent Street, Brighton, 1, Sussex.
Hawker Siddeley Aviation, Greengate, Middleton, Manchester.
Hawker Siddeley Dynamics, Manor Road, Hatfield, Herts.
Hawker Siddeley Dynamics, Industrial Autom. Div., Richmond Road, Kingston, Surrey.

Hewlett-Packard Ltd., 224 Bath Road, Slough, Bucks.
Hewlett-Packard Ltd., Dallas Road, Bedford.
Honeywell Controls Ltd., Great West Road, Brentford, Middsx.
Honeywell Controls, Hemel Hemptsead, Herts.
T. C. Howden & Co., Althorpe Street, Leaminton Spa, Warwks.
Hunting Engineering Ltd., Electrocontrols Division, Dallas Road, Bedford, Beds.
I.D.M. Electronics, Arkwright Road, Reading, Berks.
Instron Ltd., Coronation Road, High Wycombe, Bucks.
Integrated Electronics, Tilly's Lane, High Street, Staines, Middsx.
Intercole Systems Ltd., 14 Goldcroft, Yeovil, Somerset.
Intercole Systems, Portsmouth Road, Trading Estate, Ashley Crescent, Southampton.
Intermeasure Ltd., 129 London Road, Camberley, Surrey.
Kistler Instruments Ltd., The Ridges, 2 Clockhouse Road, Farnborough, Hants.
Kyowa Electronic Instruments, 19 Shiba Nishikubo, Akefune-cho, Minato-Ku, Tokyo, Japan.
Leland Instruments Ltd., 145 Grosvenor Road, London, S.W.1.
Leeds & Northrup, Wharfdale Road, Tyseley, Birmingham, 11.
Malcolm Cross Engineering (Nott'm) Ltd., Goss Street, Beeston, Nottingham.
Marconi Instruments, Longacres, St. Albans, Herts.
MB Metals Ltd., Vale Road, Portslade, Sussex.
Measurement Research, Waterloo Works, Gorsey Mount Street, Stockport, Cheshire.
M.E.L. Equipment Co. Ltd., Manor Royal, Crawley, Sussex.
Model & Prototype Systems Ltd., Kingston-on-Thames, Surrey.
Modern Precision Engineers (Finchley) Ltd., Dollis Park, London, N.3.
Mycalex & T.I.M. Ltd., Ashcroft Road, Cirencester, Glos.
Normalair-Garrett, Yeovil, Somerset.
Parametron (Stroud) Ltd., 70 Westward Road, Stroud, Glos.
Perivale Controls Co. Ltd., Civil Engineering Instrumentation Div., East Acton, London, W.3.
Plessey Co. (Components Group) Ltd., Kembrey Street, Swindon, Wilts.
The Plessey Company Ltd., Electronics Group, Ilford, Essex.
Pye Unicam, York Street, Cambridge.
Redifon Astroda Ltd., Brookside Avenue, Rustington, Sussex.
Remington Rand Ltd., Remmington House, 65 Holborn Viaduct, London, E.C.1.
Richardsons Westgarth Ltd., P.O. Box 7, Oakesway, West Hartlepool, Co. Durham.
Rocke International (UK) Limited, Tetra Pak House, Orchard Road, Richmond, Surrey.
Salford Electrical Instruments Ltd., Times Mill, Heywood, Lancs.
Sangamo Controls, North Bersted, Bognor Regis, Sussex.
Satchwell Controls Ltd., Slough, Bucks.
Savage & Parsons, Otterspool, Watford, Herts.
James Scott (Electronic Engineering) Ltd., Cartyne Industrial Estate, Glasgow, E.2.
S.E. Laboratories, North Feltham Trading Estate, Feltham, Middsx.

Serck Controls Ltd., Queensway, Leamington Spa, Warwicks.
D. Shackman & Sons, Chiltern Works, Waterside, Chesham, Bucks.
Smail Sons & Co., 21–23 India Street, Glasgow.
Smiths Industries Ltd., Industrial Instrument Div., Kelvin House, Wembley Park Drive, Wembley, Middsx.
Solartron Electronic Group Ltd., Farnborough, Hants.
Southern Instruments Ltd., Frimley Road, Camberley, Surrey.
Standard Telephones & Cables Ltd., STC House, 190 Strand, London, W.C.2.
Strainstall, Harleco House, Denmark Road, Cowes, Isle of Wight.
Techmation Ltd., 58 Edgware Way, Edgware, Middsx.
Tecquipment, Clinton House, 2A Sherwood Rise, Nottingham.
Tektronix (U.K.) Ltd., Beaverton House, Station Approach, Harpenden, Herts.
T.E.M. Sales Ltd., Gatwick Road, Crawley, Sussex.
Tensometer Ltd., Croydon, Surrey.
H. W. Tinsley, Werndee Hall, South Norwood, London, S.E.25.
Vibro-Meter Corporation, Haletop Civic Centre, Wythenshawe, Manchester, 22.
Wayne Kerr Co., New Malden, Surrey.
Welwyn Electric Strain Measurement Division, 70 High Street, Teddington, Middsx.
Westinghouse Brake & Signal Co. Ltd., 82 York Way, Kings Cross, London, N.1.
World Automation, Inc., 123 Pall Mall, London, S.W.1.

CENTRES FROM WHICH ADVICE ON STRAIN GAUGES MAY BE OBTAINED

Automation Industries U.K. Limited, Albert Street, Fleet, Aldershot, Hants., U.K. (L. Joyce).
British Society for Strain Measurement, 281, Heaton Road, Newcastle-upon-Tyne, NE6 5QB, U.K.
The City of Birmingham Polytechnic, Science and Technology South Centre, Bristol Road South, Birmingham 31, U.K. (E. J. Hearn).
Strainstall Ltd., Harleco House, Denmark Road, Cowes, I.O.W., U.K. (F. Hartshorne).
Testwell Ltd., 12, High March, Long March Industrial Estate, Daventry, Northants., U.K. (R. Ryman).
Welwyn Electric, Strain Measurement Division, 70, High Street, Teddington Middlesex, U.K. (A. L. Window).
City University St. John Street, London E.C.1., U.K. (Dr. J. A. Coutts).
and most university engineering departments.

INDEX

A.C. amplifier, 55
    carrier frequency system, 54
    systems, 54
Acoustic gauge, 59
Adhesives, 32

Balanced bridge circuit, 6
Battery effect, 55
Brimor cement, 40
Bushes, materials for, 47

Capacitance, 23, 56
    effects, 57
    gauge, 62
Capacitive balancing, 57
Carrier materials, 19
Ceramic cements, 40
Ceramics, gauges on, 38
Coefficient of expansion, 23
Concrete, gauges on, 38
Contact potentials, 55
Cost, 20
Cross sensitivity, 8

Data logger, 58
D.C. amplifier, 55
    systems, 54
Delta rosette gauge, 13
Demec gauge, 59
Differential expansion, 23
    transformer, 62
Dynamic recording, 57
Dummy gauge, 5
Duration of strain, 18

Excitation current, 37

Fatigue life, 21
Fatigue-life gauge, 16
Flexagauge, 16
Foil gauges, 8, 12

Folded filament gauge, 42
Frequency of strain, 18
Friction gauges, 15
Full-bridge system, 51

Gauge currents 38
Gauge factor, 4
    form of, 19
    length, effect of, 11
    resistance, 21
    sensitivity, 21
    size of, 19
    small, 30
Glue-line thicknesses, 27
Grid construction, 7

Half-bridge system, 5, 50
High frequency measurements, 55
High temperature measurements, 40
Hostile environments, 34
Hot spots, 38
Humidity, effect of, 34
Hysteresis, 20

Inductance, 56
    gauge, 61
Installation, 25
Instrumentation zero drift, 43
Insulation resistance, 28

Jump wires, 27

Lead wire drift, 43, 44
    resistance, 23, 24
Linear differential transformer
    strain gauge, 63
    gauges, 10
Long-term testing, 43

Materials for slip rings and
    bushes, 47

Measurement systems, basic, 49
Mechanical damage, 35
  gauge, 59
Mersenne's Law, 59
Metalfilm gauges, 23
Moisture, effect of, 34
Multi-channel strain indicator, 62

Null balance circuit, 6

Operating temperature, 19
Oscilloscope, 58

P.350 portable strain indicator, 53
Pen and ink recorder, 58
Phase balancing, 55
Physical size, 19
Piezo-resistive
  effect, 60
  gauge, 60
Plastics, gauges on, 38
Pneumatic gauge, 59
Postage stamp gauges, 14
Protection, 25

Quarter bridge system, 49

Recording systems, 58
Rectangular rosette gauge, 13
Resistivity, 3
Rokide process, 40
Rosette gauges, 12
Rotating components, measurements from, 46

Screened cables, 56
Self-adhesive gauges, 15
Semiconductor gauge, 60
Siemens three-lead wiring method, 24

Sine-wave excitation, 56
Slip rings, 46
  materials for, 47
Specific resistance, 3
Square-wave excitation, 56
Stacked rectangular rosette
  gauges, 13
Strain
  gauge, cements, 32
  drift, 43, 44
  limits, 20
Stress gauge, 16
Surface preparation procedure, 26

Telemetry, 46
Temperature coefficient of resistance, 22
  compensation, 22
Terminal strips, 27
Testing the installation, 28
Thermocouple effect, 55
Thermoelectric voltages, 55
Torque gauges, 13
Transverse sensitivity, 8, 19

Ultra-violet recorder, 58
Unbalanced bridge circuit, 5

Vibrating wire gauge, 59

Water-proofing, 35
Weldable strain gauge, 31
Welding, 42
Wheatstone bridge, 5
Wire gauges, 8

Zero load stability, 37
Zero shift, 37
  tests, 20